The Scientific Revolution: Unveiling the Secrets of the Past

Jafferson

Title: The Scientific Revolution: Unveiling the Secrets of the Past

Author's: Jefferson

This book was printed and published by [Publisher's: Jefferson] in [2023]

ISBN:

TABLE OF CONTENT

Chapter 1: The Roots of Scientific Inquiry

The Ancient World: Early Observations and Natural Philosophy

In the vast realm of intellectual history, one cannot overlook the contributions of the ancient world in shaping our understanding of the natural world. This subchapter delves into the early observations and natural philosophy of ancient civilizations, shedding light on the seeds of scientific thought that would eventually blossom into the Scientific Revolution.

The ancient world was a landscape of remarkable civilizations - from the Egyptians and Mesopotamians to the Greeks and Romans. These cultures, though separated by time and geography, shared a common curiosity about the natural world and sought to unravel its mysteries.

The Egyptians, known for their advanced understanding of astronomy, observed the celestial bodies and developed the first solar calendars. Their fascination with the stars allowed them to predict the annual flooding of the Nile, an event crucial to their agricultural practices. This early form of empirical observation laid the foundation for future scientific endeavors.

Similarly, the Mesopotamians, with their advanced mathematical knowledge, made significant contributions to the study of astronomy. They created intricate star catalogs and were the first to identify planetary movements. By meticulously recording celestial events, they established the basis for the study of celestial mechanics.

However, it was the ancient Greeks who made extraordinary strides in natural philosophy. The philosophers of ancient Greece, including

Thales, Pythagoras, and Aristotle, sought to explain the natural world through reason and observation. Thales, often referred to as the father of Western philosophy, theorized that water was the fundamental substance from which all things derived, while Pythagoras introduced the concept of mathematical relationships in explaining natural phenomena.

Aristotle, perhaps the most influential of the ancient philosophers, developed a systematic approach to scientific inquiry. He classified and categorized various natural phenomena, laying the groundwork for the scientific method. Aristotle's writings on physics, biology, and astronomy influenced generations of scholars, leaving an indelible mark on scientific thought.

The ancient world's early observations and natural philosophy set the stage for the Scientific Revolution centuries later. Their empirical approach to understanding the natural world, coupled with their philosophical inquiries, laid the foundation for the systematic study and experimentation that would come to define modern science.

As students of intellectual history, it is essential to recognize and appreciate the remarkable contributions of these ancient civilizations. By understanding their early observations and natural philosophy, we gain a deeper understanding of the origins of scientific thought and the progression of human knowledge.

The Islamic Golden Age: Preserving and Advancing Knowledge

During the Islamic Golden Age, a remarkable period spanning from the 8th to the 14th centuries, the Islamic world witnessed an unprecedented flourishing of knowledge and intellectual pursuits. This era, also known as the Islamic Renaissance, was characterized by a strong emphasis on education, scientific inquiry, and the preservation of ancient Greek and Roman texts. It is an essential chapter in intellectual history that students of all disciplines should explore to gain a deeper understanding of the roots of scientific progress.

One of the most significant contributions of the Islamic Golden Age was the preservation and translation of classical works of ancient Greece and Rome. Islamic scholars recognized the value of these texts and diligently translated them into Arabic, thereby ensuring their survival and accessibility to future generations. These translated works covered a wide range of subjects, including philosophy, mathematics, astronomy, medicine, and geography, among others. Students today owe a great debt to these scholars, as their efforts laid the foundation for the scientific revolution that would later take place in Europe.

Furthermore, Islamic scholars made substantial advancements in various scientific fields during this period. Mathematics flourished under the influence of scholars like Al-Khwarizmi, who introduced the decimal numbering system and algebra. Astronomy was revolutionized by the works of Ibn al-Haytham, whose experimental approach laid the groundwork for modern scientific methods. Medicine also saw remarkable progress, thanks to figures like Ibn Sina (Avicenna), whose comprehensive medical encyclopedias became standard references in European universities for centuries.

The Islamic Golden Age was not limited to scientific pursuits alone. It was a time of great cultural exchange, as scholars from different regions, ethnicities, and religious backgrounds came together to share knowledge and ideas. This cosmopolitan atmosphere fostered an environment of intellectual curiosity and innovation. Islamic societies also made significant advancements in architecture, literature, music, and art, leaving an indelible mark on human civilization.

For students interested in intellectual history, exploring the Islamic Golden Age unveils a rich tapestry of knowledge, innovation, and cultural exchange. It sheds light on the interconnectedness of different civilizations, challenging the notion of a singular narrative of scientific progress. By studying this era, students can gain a deeper appreciation for the contributions of Islamic scholars and understand the complex dynamics that shaped the scientific revolution.

In conclusion, the Islamic Golden Age stands as a testament to the power of knowledge and intellectual curiosity. By preserving and advancing knowledge, Islamic scholars laid the groundwork for scientific progress and contributed to the intellectual history of humanity. Students today should embrace the opportunity to study this pivotal era, as it provides valuable insights into the roots of scientific development and the importance of cultural exchange in shaping our world.

The Renaissance: Rediscovering Ancient Wisdom

The Renaissance, a period of profound cultural and intellectual transformation spanning roughly from the 14th to the 17th century, marked a pivotal chapter in history. This subchapter, titled "The Renaissance: Rediscovering Ancient Wisdom," delves into the essence of this era and explores its impact on intellectual history.

During the Renaissance, Europe experienced a remarkable resurgence of interest in the classical knowledge of ancient civilizations, particularly those of Greece and Rome. This renewed fascination with the wisdom of the past had a profound influence on various fields of study, including philosophy, art, literature, and science. It was a time of great intellectual curiosity and exploration, leading to numerous groundbreaking discoveries and advancements.

One of the central aspects of the Renaissance was the revival of humanism, a philosophy that emphasized the importance of human potential, individualism, and the pursuit of knowledge. Humanists sought to understand and embrace the teachings of ancient scholars, such as Aristotle, Plato, and Cicero. They emphasized the importance of critical thinking and rational inquiry, and this approach laid the foundation for much of the intellectual progress that followed.

The Renaissance also witnessed the birth of the printing press, a revolutionary invention that allowed for the mass production and dissemination of knowledge. As a result, ancient texts and scholarly works became more accessible to a wider audience, fueling the spread of intellectual ideas throughout Europe. This newfound availability of information played a crucial role in inspiring new thinkers and fostering intellectual dialogue.

Perhaps one of the most influential figures of the Renaissance was Leonardo da Vinci, an unparalleled polymath whose curiosity knew no bounds. His insatiable thirst for knowledge and relentless pursuit of truth epitomized the spirit of the era. Da Vinci's contributions spanned across various disciplines, from art and architecture to anatomy and engineering, and his work exemplified the synthesis of ancient wisdom and innovative thinking.

The Renaissance not only revived ancient wisdom but also challenged and reinterpreted it. Scholars of the period sought to reconcile classical teachings with the Christian faith, leading to the emergence of new ideas and perspectives. This intellectual synthesis paved the way for the Scientific Revolution, a later period of unprecedented advancements in scientific discovery.

In conclusion, the Renaissance was a transformative period that witnessed the rediscovery and reevaluation of ancient wisdom. Its impact on intellectual history was profound, shaping the way we perceive knowledge, reason, and human potential. By embracing the teachings of the past, the Renaissance laid the groundwork for the scientific and intellectual progress that would follow, leaving an indelible mark on the course of human history.

Chapter 2: The Predecessors of the Scientific Revolution

Copernicus and the Heliocentric Model

In the vast expanse of intellectual history, there are few figures whose ideas have had such a profound impact on our understanding of the world as Nicolaus Copernicus. Born in 1473 in what is now Poland, Copernicus was a mathematician, astronomer, and Catholic cleric whose groundbreaking work challenged the prevailing geocentric model of the universe and set the stage for the scientific revolution.

At the time, the dominant belief was that Earth stood at the center of the cosmos, with the sun, moon, and planets revolving around it. This geocentric model had its roots in ancient Greek philosophy and was further solidified by the influential work of Ptolemy in the 2nd century AD. However, Copernicus dared to question this long-standing orthodoxy.

In his seminal work, "De Revolutionibus Orbium Coelestium" (On the Revolutions of the Celestial Spheres), published in 1543, Copernicus put forth a revolutionary idea – the heliocentric model. According to this model, the sun, not the Earth, was at the center of the universe, and the planets, including Earth, orbited around it in perfect circles. This concept challenged not only the prevailing scientific understanding but also the religious and philosophical doctrines of the time.

Copernicus's heliocentric model was not without its flaws. For instance, he still clung to the notion of circular orbits, which did not accurately predict the observed planetary motions. However, his work laid the foundation for future scientists, such as Johannes Kepler and Isaac Newton, to refine and expand upon his ideas, ultimately leading to a more accurate understanding of celestial mechanics.

The impact of Copernicus's heliocentric model cannot be overstated. It not only revolutionized our understanding of the cosmos but also sparked a fundamental shift in the way we approach scientific inquiry. By challenging long-held beliefs, Copernicus set the stage for a new era of intellectual curiosity and empirical investigation that would shape the scientific revolution.

For students of intellectual history, studying Copernicus and the heliocentric model offers a fascinating window into the development of scientific thought. It highlights the power of human curiosity and the courage to challenge established paradigms. Copernicus's work serves as a reminder that knowledge is never static but rather constantly evolving as new evidence emerges and old ideas are discarded.

In conclusion, Copernicus's heliocentric model stands as a testament to the power of intellectual curiosity and the importance of questioning prevailing beliefs. His groundbreaking work not only reshaped our understanding of the cosmos but also paved the way for the scientific revolution that would forever change the course of human history.

Brahe and Kepler: Advancing Planetary Motion

In the vast realm of intellectual history, few names shine as brightly as those of Tycho Brahe and Johannes Kepler. These two remarkable individuals played a crucial role in advancing our understanding of planetary motion during the Scientific Revolution, forever changing our perception of the cosmos.

Tycho Brahe, a Danish nobleman and astronomer, was known for his meticulous observations of the night sky. Armed with innovative instruments, Brahe meticulously recorded the positions of celestial bodies, amassing a treasure trove of data that would prove invaluable to future astronomers. Brahe's observations challenged the prevailing Aristotelian model of the universe, which held that celestial bodies moved in perfect circles. His careful observations revealed inconsistencies in the motion of planets, paving the way for a new understanding of the cosmos.

Enter Johannes Kepler, a German mathematician and astronomer deeply influenced by Brahe's work. Kepler was the first to recognize the true nature of planetary motion, eventually formulating his famous three laws of planetary motion. Kepler demonstrated that the planets moved in elliptical orbits, with the Sun at one of the focal points. This revolutionary insight shattered the long-held belief in circular orbits and provided a more accurate description of planetary motion.

Kepler's laws not only explained the motion of planets but also provided a mathematical framework for astronomers to predict their future positions. This development was a breakthrough for celestial navigation and laid the foundation for future astronomers to explore the universe with greater precision.

The collaboration between Brahe and Kepler was a turning point in the Scientific Revolution. Brahe's meticulous observations and Kepler's mathematical genius combined to produce a paradigm shift in our understanding of the cosmos. Their groundbreaking work challenged ancient beliefs and set the stage for the birth of modern astronomy.

For students delving into intellectual history, Brahe and Kepler's story serves as a testament to the power of observation and mathematical reasoning in the pursuit of scientific knowledge. Their work not only advanced our understanding of planetary motion but also paved the way for future scientific discoveries that continue to shape our world today.

As we explore the secrets of the past, it is essential to recognize the contributions of these two remarkable individuals. Brahe and Kepler's collaboration stands as a shining example of how science and intellectual curiosity can converge, transforming our understanding of the universe and propelling humanity forward on the path of knowledge.

Galileo and the Birth of Experimental Science

In the realm of intellectual history, few figures have left a more profound impact than Galileo Galilei. He is often regarded as the father of modern science and his groundbreaking experiments laid the foundation for the scientific method we use today. Galileo's contributions not only revolutionized our understanding of the physical world but also challenged the prevailing dogmas of his time, paving the way for a new era of scientific inquiry.

Born in 1564 in Pisa, Italy, Galileo displayed an innate curiosity and an insatiable thirst for knowledge from an early age. His passion for mathematics and physics led him to the University of Pisa, where he excelled in his studies and began questioning the prevailing Aristotelian worldview that dominated academia at the time. Galileo's genius lay in his ability to merge theoretical knowledge with practical experiments, an approach that would come to define the birth of experimental science.

One of Galileo's most famous experiments involved the study of motion. By rolling balls of different weights down inclined planes, he discovered that the time it took for each ball to reach the bottom was independent of its weight. This contradicted Aristotle's belief that heavier objects fall faster than lighter ones. Galileo's meticulous observations and measurements provided empirical evidence that challenged centuries-old theories, sparking a revolution in scientific thought.

Galileo's pioneering work extended to the field of astronomy as well. Armed with a newly invented telescope, he made groundbreaking observations of the heavens. Galileo discovered the four largest moons of Jupiter, providing evidence that not everything in the cosmos revolved around the Earth. This challenged the geocentric model of the universe, which held that Earth was at the center and all celestial bodies orbited around it. Galileo's findings supported the heliocentric model proposed by Copernicus, further undermining traditional beliefs and setting the stage for a paradigm shift.

However, Galileo's revolutionary ideas and his challenge to the authority of the Catholic Church attracted controversy and eventually led to his persecution. In 1633, he was tried by the Inquisition and forced to recant his heliocentric views. Despite his personal setback, Galileo's legacy endured, inspiring generations of scientists to question established truths and explore the world through experimentation.

Galileo's contributions to the birth of experimental science cannot be overstated. His relentless pursuit of knowledge, his insistence on empirical evidence, and his courage in challenging conventional wisdom laid the groundwork for the scientific revolution that would follow. His story serves as a powerful reminder to students of the immense power of curiosity, critical thinking, and the pursuit of truth.

Chapter 3: The Breakthroughs of the Scientific Revolution

Francis Bacon and the Scientific Method

In the realm of intellectual history, few figures have made a greater impact than Sir Francis Bacon. Known as the father of empiricism and the scientific method, Bacon's contributions to the field of science laid the foundation for the modern scientific revolution. His ideas not only challenged the prevailing beliefs of his time but also shaped the way we understand and approach scientific inquiry today.

Born in 1561, Bacon's early education and legal career provided him with a solid foundation for his later philosophical pursuits. However, it was his keen interest in natural philosophy and his dissatisfaction with the prevailing scholastic approach that set him apart. Bacon believed that knowledge should be acquired through observation and experimentation, rather than relying solely on deductive reasoning or relying on the authority of ancient texts.

Bacon's most notable work, Novum Organum, published in 1620, presented a new method of scientific inquiry. He advocated for a systematic approach that combined inductive reasoning, careful observation, and experimentation. Bacon argued that scientific knowledge should be derived from the careful collection and analysis of data, which he called "natural history." This data would then be subjected to rigorous experimentation and analysis to uncover the true nature of phenomena.

One of Bacon's key contributions was his emphasis on the importance of collaboration and the sharing of knowledge. He believed that scientific progress could only be achieved through collective efforts, and he advocated for the establishment of scientific societies and institutions. Bacon's vision of a scientific community where information was freely exchanged and ideas were tested and refined laid the groundwork for the development of modern scientific societies.

Bacon's ideas were revolutionary for his time and challenged the entrenched beliefs of the era. His emphasis on empirical observation, experimentation, and collaboration paved the way for the scientific method that is widely used today. By separating science from philosophy and religion, Bacon helped establish the autonomy of scientific inquiry, encouraging scholars to explore the natural world with an open mind and a spirit of curiosity.

In conclusion, Francis Bacon's contributions to the scientific method and his emphasis on empirical observation and experimentation had a profound impact on the intellectual history of the scientific revolution. His ideas continue to shape the way we approach scientific inquiry and are a testament to the power of human curiosity and the pursuit of knowledge. By understanding Bacon's legacy, students of intellectual history can gain valuable insights into the foundations of modern science and the importance of evidence-based inquiry.

Descartes and the Role of Reason

In the realm of intellectual history, René Descartes stands as a towering figure, whose ideas and philosophies continue to shape our understanding of the world today. Descartes, a French philosopher, mathematician, and scientist, played a crucial role in the Scientific Revolution of the 17th century. His relentless pursuit of truth and emphasis on reason revolutionized the way we think about knowledge, skepticism, and the nature of reality.

Descartes firmly believed that reason was the key to unlocking the mysteries of the universe. He advocated for a systematic approach to knowledge, rejecting traditional beliefs and relying solely on reason and logic. This approach, known as rationalism, asserts that true knowledge can only be attained through the power of reason, independent of sensory experience.

In his groundbreaking work, "Meditations on First Philosophy," Descartes embarked on a quest to establish a foundation of knowledge that could withstand doubt and skepticism. He famously employed the method of doubt, systematically doubting everything he had previously believed to be true. By stripping away all preconceived notions and beliefs, Descartes aimed to find a solid and indubitable foundation for knowledge.

Descartes arrived at his famous maxim, "I think, therefore I am," as a result of this rigorous process of doubt. He argued that even if all of our senses deceive us, the mere act of doubting implies the existence of a doubter. From this point, Descartes built his entire philosophy on the foundation of reason and the certainty of one's own existence.

Descartes' philosophy extended beyond metaphysics and epistemology, and he made significant contributions to the fields of mathematics and science. He developed a groundbreaking system of analytical geometry, which laid the foundation for modern algebraic geometry. His works on optics and the laws of motion also influenced the development of classical physics.

Descartes' emphasis on reason and skepticism had a profound impact on the Scientific Revolution and subsequent intellectual history. His ideas challenged the prevailing beliefs of his time and paved the way for a new era of scientific inquiry. Descartes' emphasis on reason as the ultimate arbiter of truth continues to resonate with thinkers and scholars to this day.

In conclusion, Descartes' role in intellectual history cannot be overstated. His relentless pursuit of truth and emphasis on reason transformed our understanding of knowledge and reality. By challenging traditional beliefs and advocating for a systematic approach to knowledge, Descartes laid the groundwork for the Scientific Revolution and shaped the course of intellectual history. Students of intellectual history can find inspiration in Descartes' unwavering commitment to reason and his enduring impact on the development of scientific inquiry.

Newton and the Laws of Motion

In the realm of intellectual history, few figures have had as profound an impact on science as Sir Isaac Newton. His groundbreaking work on the laws of motion laid the foundation for classical mechanics and revolutionized our understanding of the physical world. In this subchapter, we delve into Newton's life, his laws of motion, and their significance in shaping the scientific revolution.

Born in 1643 in Woolsthorpe, England, Newton displayed an early aptitude for mathematics and physics. His curiosity and relentless pursuit of knowledge led him to develop three fundamental laws of motion that remain fundamental to this day. These laws provided a systematic explanation of how objects move and interact with one another, forever changing the way scientists approached the study of motion.

Newton's first law of motion, also known as the law of inertia, states that an object at rest will remain at rest, and an object in motion will continue moving in a straight line at a constant speed unless acted upon by an external force. This law challenged the prevailing belief that objects required a constant force to remain in motion, leading to a paradigm shift in scientific thinking.

The second law of motion introduced the concept of force and its relationship to mass and acceleration. It states that the force acting on an object is equal to its mass multiplied by its acceleration. This law provided a quantitative framework for understanding how forces influence an object's motion, allowing scientists to make precise predictions and calculations.

The third law of motion, often referred to as the law of action and reaction, states that for every action, there is an equal and opposite reaction. This law elucidates the reciprocal nature of forces, emphasizing that every force exerted on an object is met with an equal force exerted in the opposite direction. It laid the groundwork for understanding the dynamics of interactions between objects and the conservation of momentum.

Newton's laws of motion not only revolutionized the field of physics but also had far-reaching implications for other scientific disciplines. They provided a solid foundation for understanding the motion of planets, the behavior of fluids, and even the principles behind engineering and architecture.

In conclusion, Newton's laws of motion stand as a testament to his genius and mark a turning point in the history of science. By unveiling the secrets of motion, Newton paved the way for future scientific advancements and laid the groundwork for the modern scientific method. As students of intellectual history, it is crucial to recognize and appreciate the profound impact of Newton's laws, as they continue to shape our understanding of the physical world around us.

Chapter 4: The Revolution in Astronomy and Cosmology

Kepler's Laws and the Shape of Planetary Orbits

In the vast realm of intellectual history, few individuals have left a more profound impact than Johannes Kepler, a key figure in the Scientific Revolution. Kepler's groundbreaking work on the laws governing planetary motion revolutionized our understanding of the universe and paved the way for modern astronomy. In this subchapter, we delve into Kepler's Laws and explore how they shaped our understanding of the shape of planetary orbits.

Kepler's first law, also known as the law of ellipses, challenged the prevailing belief that planetary orbits were perfect circles. Through meticulous observations of Mars, Kepler discovered that planets follow elliptical paths around the Sun, with the Sun occupying one of the foci of the ellipse. This finding shattered the age-old notion of circular celestial motion and revealed the true nature of planetary orbits.

The second law, known as the law of areas, deals with the speed at which a planet moves along its elliptical path. Kepler observed that a planet sweeps out equal areas in equal time periods, regardless of its position in the orbit. This law implies that a planet moves faster when it is closer to the Sun and slower when it is farther away, a concept that was revolutionary at the time. This insight offered a new perspective on the dynamics of celestial bodies and contributed to the development of Newton's laws of motion.

The third law, often referred to as the law of harmonies, established a mathematical relationship between the orbital period of a planet and its distance from the Sun. Kepler found that the square of a planet's orbital period is proportional to the cube of its average distance from the Sun. This discovery provided a quantitative framework for understanding the vast range of planetary motion in our solar system.

Kepler's Laws not only transformed our understanding of planetary motion but also laid the foundation for future scientific advancements. They challenged long-held beliefs, revolutionized the field of astronomy, and paved the way for Newton's laws of motion. Kepler's work remains a testament to the power of observation, mathematical reasoning, and the pursuit of knowledge.

As students of intellectual history, it is crucial to appreciate the impact of Kepler's Laws on our current understanding of the universe. By studying the scientific breakthroughs of the past, we gain a deeper appreciation for the scientific method, the importance of empirical evidence, and the ongoing quest to unravel the secrets of the cosmos. Kepler's Laws serve as a reminder that the pursuit of knowledge knows no bounds and that great discoveries lie ahead for those who dare to question and explore.

Galileo's Observations and the Confirmation of Heliocentrism

One of the most pivotal moments in the history of science can be attributed to the observations and discoveries made by the renowned Italian astronomer, Galileo Galilei. His groundbreaking work not only challenged the prevailing belief in a geocentric universe but also provided substantial evidence to support the heliocentric model proposed by Nicolaus Copernicus.

In the early 17th century, the notion that the Earth was at the center of the universe was deeply ingrained in Western intellectual thought. However, Galileo's meticulous observations through his newly invented telescope began to unravel this long-held belief. Through his telescope, Galileo discovered a multitude of celestial bodies, including the four largest moons of Jupiter, now known as the Galilean moons. These observations directly contradicted the geocentric model, which held that all celestial bodies revolved around the Earth.

Galileo's observations of the phases of Venus were particularly significant. By observing Venus at different points in its orbit around the Sun, he noticed that it exhibited phases similar to those of the Moon. This observation was only possible if Venus orbited the Sun, providing strong evidence for the heliocentric model.

Despite the compelling evidence he presented, Galileo faced considerable opposition from the Catholic Church, which staunchly adhered to the geocentric worldview. In 1616, the Church issued a decree that declared Copernicanism as "false and contrary to Scripture." Galileo was forbidden from teaching or defending these ideas.

Undeterred, Galileo continued his scientific pursuits and published his seminal work, "Dialogue Concerning the Two Chief World Systems," in 1632. This book presented a dialogue between three characters, one of whom argued in favor of the heliocentric model. Although Galileo claimed that his work was neutral and merely presented arguments from both sides, it was seen as a direct challenge to the Church's teachings.

As a result, Galileo was summoned before the Inquisition in 1633 and compelled to renounce his views. He spent the rest of his life under house arrest, but his ideas had already spread far and wide, leaving an indelible impact on the intellectual history of the scientific revolution.

Galileo's observations and unwavering belief in the heliocentric model laid the foundation for future scientific breakthroughs. His work paved the way for astronomers and scientists to challenge existing beliefs, encouraging a more empirical and evidence-based approach to scientific inquiry. Galileo's legacy serves as a reminder that scientific progress often requires questioning long-standing dogmas and embracing the pursuit of knowledge, even in the face of opposition.

Newton's Universal Law of Gravitation

In the vast realm of intellectual history, Sir Isaac Newton stands as a towering figure whose groundbreaking discoveries revolutionized our understanding of the universe. One of his most remarkable contributions is the development of the Universal Law of Gravitation, a concept that continues to shape our understanding of the cosmos to this day.

Newton's Universal Law of Gravitation emerged during the Scientific Revolution, a period of intense scientific inquiry and discovery spanning from the 16th to the 18th century. At the time, prevailing beliefs held that objects fell to the ground due to some inherent force within them. However, Newton challenged this notion by proposing that the force responsible for this phenomenon was not contained within the objects themselves, but rather existed between all objects in the universe.

According to Newton's law, every particle of matter in the universe attracts every other particle with a force that is directly proportional to the product of their masses and inversely proportional to the square of the distance between them. This means that the force of gravity acting on an object increases as its mass increases, and decreases as the distance between objects increases.

The implications of Newton's Universal Law of Gravitation were profound. It provided a mathematical framework to explain not only the motions of objects on Earth but also the movements of celestial bodies in space. For the first time, the laws governing the motion of objects on Earth and in the heavens were unified under a single set of principles.

Newton's law also allowed scientists to make accurate predictions about the motion of planets and moons, leading to a deeper understanding of the solar system. His calculations enabled the prediction of astronomical events such as eclipses and the motion of comets, which were previously shrouded in mystery.

Furthermore, Newton's Universal Law of Gravitation laid the foundation for the development of classical mechanics, a branch of physics that describes the motion of objects under the influence of forces. This groundbreaking concept paved the way for future scientific advancements and influenced subsequent generations of scientists, including Albert Einstein and his theory of general relativity.

In conclusion, Newton's Universal Law of Gravitation is a cornerstone of intellectual history. It not only revolutionized our understanding of gravity but also paved the way for a deeper comprehension of the cosmos. By unifying the laws of motion on Earth and in space, Newton's law forever changed the course of scientific inquiry and continues to inspire students and scientists alike to unveil the secrets of the past.

Chapter 5: The Revolution in Medicine and Anatomy

Vesalius and the Study of Human Anatomy

In the realm of intellectual history, few individuals have left a lasting impact like Andreas Vesalius. Born in 1514 in Brussels, Vesalius revolutionized the study of human anatomy during the Scientific Revolution. His groundbreaking work not only challenged long-held beliefs but also laid the foundation for modern medicine and our understanding of the human body.

During the Renaissance, the study of anatomy was primarily based on the works of the ancient Greek physician Galen. However, Vesalius, dissatisfied with the inaccuracies and limitations of Galenic anatomy, embarked on a mission to examine the human body firsthand. As a result, he published his magnum opus, "De humani corporis fabrica" (On the Fabric of the Human Body), in 1543.

Vesalius's meticulous dissections and observations led to significant discoveries that overturned many of Galen's teachings. He emphasized the importance of direct observation, challenging the prevailing belief that knowledge could only be gained through the interpretation of ancient texts. Vesalius's work demonstrated the necessity of examining the human body itself to understand its intricate structure and function.

In his book, Vesalius presented detailed illustrations of the human body, depicting its various systems and organs. These anatomical drawings were not only accurate but also aesthetically pleasing, making them invaluable resources for students and scholars alike.

Vesalius's emphasis on visual representation transformed the way anatomy was taught and studied, paving the way for future anatomists and medical illustrators.

Moreover, Vesalius's work had profound implications for medicine. His precise descriptions and classifications of anatomical structures enabled more accurate diagnoses and surgical procedures. Vesalius's emphasis on the importance of anatomy as a foundation for medical practice improved patient care and set new standards for medical education.

Vesalius's contributions to the study of human anatomy exemplify the spirit of the Scientific Revolution. His dedication to empirical observation, critical thinking, and the pursuit of knowledge challenged the prevailing dogmas of his time. By breaking away from traditional beliefs and relying on direct evidence, Vesalius shattered the barriers that had confined the study of human anatomy for centuries.

Today, Vesalius's legacy lives on. His work continues to inspire students and scholars in the field of medicine and intellectual history. The Scientific Revolution owes much of its progress to Vesalius's unwavering commitment to truth and his unwavering pursuit of knowledge.

Harvey and the Circulation of Blood

In the fascinating realm of intellectual history, there are few individuals who have left an indelible mark on the course of scientific progress quite like William Harvey. His groundbreaking discoveries in the field of medicine revolutionized our understanding of the human body and forever changed the way we perceive the circulation of blood.

During the 17th century, the prevailing belief was that blood was produced in the liver and consumed by the body as fuel. It was commonly thought that blood flowed back and forth between the heart and the liver, with no clear understanding of how it circulated throughout the rest of the body. This prevailing view held sway for centuries until Harvey came along with his meticulous observations and astute reasoning.

Harvey, a brilliant English physician, challenged the existing theories and embarked on a quest to uncover the secrets of blood circulation. Through dissections, careful observations, and experimentation, he proposed a groundbreaking concept—that the heart acted as a pump, propelling blood throughout the body in a continuous loop.

In 1628, Harvey published his magnum opus, "Exercitatio Anatomica de Motu Cordis et Sanguinis in Animalibus" (An Anatomical Exercise on the Motion of the Heart and Blood in Animals). In this seminal work, he presented a detailed account of the heart's function and the circulation of blood. Harvey demonstrated that the heart served as a muscular pump, pumping oxygenated blood to the rest of the body through the arteries, while the veins carried deoxygenated blood back to the heart.

Harvey's meticulous observations and experiments laid the foundation for modern physiology and our understanding of the circulatory system. His work challenged long-held notions and paved the way for further scientific inquiry, setting a new standard for evidence-based medicine.

The impact of Harvey's discoveries extended far beyond the field of medicine. His work influenced the entire scientific community, sparking a revolution in the way knowledge was pursued and validated. By embracing empirical evidence and critical thinking, Harvey exemplified the spirit of the Scientific Revolution, which would shape the course of intellectual history for centuries to come.

For students of intellectual history, studying the life and work of William Harvey is a window into the profound shifts that occurred during the Scientific Revolution. His relentless pursuit of knowledge, meticulous observations, and groundbreaking conclusions serve as an inspiration for those seeking to understand the secrets of the past and unravel the mysteries of the natural world. Harvey's legacy reminds us of the transformative power of scientific inquiry and the enduring impact it can have on our understanding of ourselves and the universe we inhabit.

Jenner and the Discovery of Vaccination

In the realm of intellectual history, few figures have had such a profound impact on public health as Edward Jenner, the pioneer behind the discovery of vaccination. His groundbreaking work not only revolutionized medicine but also paved the way for the eradication of numerous deadly diseases. This subchapter will delve into Jenner's life, his momentous discovery, and the enduring legacy of vaccination.

Born in 1749 in rural England, Jenner grew up with a keen interest in natural history and medicine. As a country doctor, he observed that milkmaids who contracted cowpox seemed to be immune to the more severe smallpox. This observation sparked Jenner's curiosity and led him on a quest to uncover the secrets of immunity.

Jenner's meticulous research and experimentation eventually led him to develop the first vaccine. In 1796, he successfully inoculated James Phipps, an eight-year-old boy, with cowpox material, demonstrating the protective effect it had against smallpox. This breakthrough marked the birth of vaccination, a term derived from the Latin word "vacca" for cow.

Jenner's discovery faced initial skepticism and resistance from the medical community. However, his persistence and unwavering belief in the potential of vaccination eventually triumphed. His findings were published in 1798, and by the early 19th century, vaccination was widely accepted as a pivotal tool in preventing infectious diseases.

The impact of Jenner's work cannot be overstated. Smallpox, once a devastating scourge that claimed countless lives, was eventually eradicated, thanks to the widespread adoption of vaccination. His research laid the foundation for the development of vaccines against a multitude of diseases, leading to the prevention of illnesses such as polio, measles, and hepatitis.

Moreover, Jenner's discovery sparked a revolution in public health practices. Vaccination campaigns were launched worldwide, saving millions of lives and leading to significant improvements in global health. The establishment of vaccination programs became a cornerstone of disease prevention and control, shaping the field of epidemiology.

Today, vaccination continues to be a critical component of public health strategies, safeguarding individuals and communities from preventable diseases. Jenner's remarkable contribution to medicine and his enduring legacy serve as an inspiration to future generations of scientists and health professionals.

In conclusion, Edward Jenner's discovery of vaccination represents a pivotal moment in history. His relentless pursuit of knowledge and his determination to alleviate human suffering revolutionized medicine and paved the way for the eradication of devastating diseases. Jenner's legacy is a testament to the power of scientific inquiry and the profound impact it can have on the world.

Chapter 6: The Revolution in Chemistry and Alchemy

Boyle and the Experimental Approach to Chemistry

In the vast realm of intellectual history, few figures have had as profound an impact on the development of scientific thought as Robert Boyle. Renowned for his groundbreaking contributions to chemistry, Boyle's experimental approach revolutionized the way we perceive and understand the natural world. This subchapter explores Boyle's pivotal role in the scientific revolution and how his methodology forever altered the course of chemistry.

At the heart of Boyle's approach was his unwavering commitment to empirical investigation. He believed that theories and ideas should be tested and validated through experiments, rather than relying solely on philosophical or theoretical speculation. Boyle's emphasis on experimentation marked a crucial departure from the prevailing Aristotelian tradition, which heavily relied on deductive reasoning and thought experiments.

One of Boyle's most significant achievements was his pioneering work on the properties of gases. Through meticulous experimentation, he discovered Boyle's Law, which states that the volume of a gas is inversely proportional to its pressure, when temperature remains constant. This groundbreaking finding laid the foundation for our modern understanding of gas behavior and had profound implications for fields such as physics and engineering.

Furthermore, Boyle's experiments with chemical reactions and substances laid the groundwork for the field of modern chemistry. He meticulously recorded his observations and meticulously conducted experiments to investigate various chemical properties. His work on elements and compounds helped establish the concept of chemical elements and their fundamental properties.

Boyle's experimental approach to chemistry not only advanced scientific knowledge but also paved the way for the development of the scientific method itself. His meticulous documentation, systematic experimentation, and emphasis on empirical evidence became the gold standard for scientific inquiry. Boyle's methodology laid the groundwork for future generations of scientists to build upon and refine his findings, ultimately propelling the scientific revolution forward.

For students of intellectual history, understanding Boyle's contributions is essential to comprehending the evolution of scientific thought. By embracing experiment and observation over speculation, Boyle challenged the prevailing paradigms of his time and ushered in a new era of scientific inquiry. His dedication to empirical evidence and his insatiable curiosity pushed the boundaries of knowledge and set the stage for countless scientific breakthroughs that followed.

In conclusion, Robert Boyle's experimental approach to chemistry revolutionized the scientific landscape of his time and left an indelible mark on intellectual history. His emphasis on empirical evidence, meticulous experimentation, and systematic documentation paved the way for future scientific advancements. Students of intellectual history will undoubtedly find inspiration in Boyle's unwavering commitment to the pursuit of knowledge and his profound impact on the scientific revolution.

Lavoisier and the Discovery of Oxygen

In the realm of intellectual history, few individuals have left a mark as profound as Antoine-Laurent de Lavoisier, particularly through his groundbreaking discovery of oxygen. Lavoisier was a French chemist and nobleman who lived during the era of the Scientific Revolution. His contributions not only revolutionized the field of chemistry but also laid the foundation for modern scientific methodology.

During the late 18th century, the understanding of chemical reactions was still in its infancy. Scientists were struggling to comprehend the nature of combustion, the process through which substances burned and released various gases. Lavoisier, armed with a brilliant mind and a rigorous scientific approach, embarked on a series of experiments that would change the course of scientific thought forever.

Through meticulous observation and meticulous record-keeping, Lavoisier was able to demonstrate that substances burned by combining with a gas that he named "oxygen." His experiments conclusively proved that this newly discovered element played a crucial role in combustion, as well as in respiration and the process of rusting. This breakthrough shattered the prevailing phlogiston theory, which posited that a substance called "phlogiston" was released during combustion.

Lavoisier's discovery of oxygen not only had a profound impact on chemistry but also on our understanding of the natural world. It paved the way for the development of modern chemical nomenclature and laid the groundwork for the concept of chemical elements. By embracing quantitative analysis and meticulous experimentation, Lavoisier established chemistry as a true scientific discipline.

Moreover, Lavoisier's work on oxygen had far-reaching implications beyond the confines of chemistry. His experiments challenged prevailing ideas about the nature of air and its role in various natural phenomena. Furthermore, his research provided important insights into the composition of the Earth's atmosphere, which profoundly influenced the emerging field of meteorology.

Lavoisier's contributions to the scientific revolution were not limited to his discovery of oxygen alone. He also made significant strides in the field of stoichiometry, establishing the law of conservation of mass. This principle stated that atoms are neither created nor destroyed in a chemical reaction, but merely rearranged. This revolutionary concept laid the foundation for modern chemistry and remains a fundamental principle in the field to this day.

In conclusion, Lavoisier's discovery of oxygen was a pivotal moment in the history of science. His meticulous experiments and rigorous scientific methodology forever changed our understanding of chemical reactions and the nature of matter. By embracing quantitative analysis and rejecting outdated theories, Lavoisier paved the way for modern chemistry and left an indelible mark on intellectual history. Students of science and intellectual history alike can draw inspiration from Lavoisier's unwavering dedication to the pursuit of knowledge and his relentless pursuit of truth.

Dalton and the Atomic Theory

In the realm of intellectual history, few individuals have had as profound an impact on the scientific community as John Dalton. His groundbreaking work on the atomic theory revolutionized our understanding of matter and laid the foundation for modern chemistry. In this subchapter, we will delve into the life and contributions of Dalton, exploring how his ideas transformed the scientific landscape.

John Dalton, born in England in 1766, was a polymath with a deep passion for the natural sciences. He dedicated his life to uncovering the secrets of matter, ultimately leading to the formulation of the atomic theory. Dalton's theory proposed that all matter is made up of tiny indivisible particles called atoms, which combine in specific ratios to form compounds. This groundbreaking concept challenged prevailing beliefs and provided a framework for understanding chemical reactions and the behavior of different substances.

One of Dalton's most significant contributions was his insight into the concept of the atom's weight and reactivity. He proposed that each element is composed of atoms with a unique mass, and that during chemical reactions, atoms rearrange to form new compounds while retaining their original weights. Dalton's meticulous experiments and calculations paved the way for the development of the periodic table and greatly enhanced our understanding of chemical reactions.

Furthermore, Dalton's atomic theory had broader implications beyond chemistry. It provided a solid foundation for the field of physics, as it explained the behavior of gases and the relationship between temperature, pressure, and volume. Dalton's work on the atomic theory also influenced subsequent generations of scientists, inspiring further research and discoveries in the realm of atomic and molecular structures.

Despite facing initial skepticism from his contemporaries, Dalton's ideas gradually gained acceptance and transformed the scientific community. His meticulous observations and experimental approach set a precedent for scientific inquiry, emphasizing the importance of empirical evidence and rigorous methodology.

In conclusion, John Dalton's atomic theory revolutionized our understanding of matter and its fundamental building blocks. His contributions to intellectual history, specifically in the fields of chemistry and physics, continue to resonate with scientists and scholars to this day. Dalton's work serves as a reminder of the power of perseverance, observation, and theoretical innovation in shaping our understanding of the natural world. By studying Dalton's achievements, students can gain valuable insights into the scientific revolution and the profound impact it had on shaping our modern world.

Chapter 7: The Revolution in Physics and Mathematics

Galileo and the Laws of Motion

In the realm of intellectual history, few figures have left as profound an impact as Galileo Galilei. Known as the father of modern observational astronomy, Galileo's contributions extended far beyond the stars. His revolutionary experiments and observations not only challenged long-held beliefs but also laid the foundation for our understanding of the laws of motion.

During the Scientific Revolution, Galileo's experiments with inclined planes and rolling balls led him to formulate the concept of inertia. He demonstrated that objects in motion tend to stay in motion unless acted upon by an external force. This revolutionary idea contradicted the prevailing Aristotelian notion that objects naturally sought a state of rest. Galileo's experiments and subsequent laws of motion became the cornerstone of Isaac Newton's later work on classical mechanics.

One of Galileo's most significant contributions was his refinement of the telescope, which enabled him to observe celestial bodies with unprecedented clarity. Through these observations, Galileo confirmed the Copernican heliocentric model, which placed the Sun at the center of the universe. This direct challenge to the geocentric model, supported by the Catholic Church, ultimately led to Galileo's trial and house arrest. Despite the personal consequences, Galileo's work paved the way for the acceptance of the heliocentric model and revolutionized our understanding of the cosmos.

Galileo's experiments and observations extended beyond astronomy. He also investigated the laws governing the motion of bodies on Earth. Galileo's studies of falling bodies led him to propose that all objects, regardless of their mass, fall at the same rate in the absence of air resistance. This concept, known as the principle of equivalence, was a precursor to Newton's law of universal gravitation.

Galileo's scientific discoveries and his unwavering commitment to empirical evidence transformed our understanding of the physical world. His ideas challenged the entrenched beliefs of his time and set the stage for the scientific method as we know it today. Galileo's legacy serves as an inspiration for students of intellectual history, highlighting the power of observation, experimentation, and the pursuit of knowledge.

In conclusion, Galileo Galilei's groundbreaking work in astronomy, mechanics, and the laws of motion redefined our understanding of the natural world. His experiments and observations shattered long-held beliefs, paving the way for the Scientific Revolution. Galileo's unwavering dedication to empirical evidence and his commitment to the pursuit of truth continue to inspire students of intellectual history to challenge established dogmas and embrace the power of scientific inquiry.

Descartes and the Development of Analytic Geometry

In the realm of intellectual history, few figures have had a more profound impact than René Descartes. Born in France in 1596, Descartes is widely regarded as one of the key figures of the Scientific Revolution. His groundbreaking work in mathematics and philosophy laid the foundation for modern analytical thinking, with one of his most significant contributions being the development of analytic geometry.

Descartes' interest in mathematics began during his time as a soldier, where he had ample opportunity to ponder and explore the mysteries of the universe. He believed that mathematics provided a language through which the secrets of nature could be deciphered. Inspired by the works of ancient mathematicians such as Euclid and Archimedes, Descartes sought to merge algebra and geometry into a unified system.

In his seminal work, "La Géométrie," published in 1637, Descartes presented his revolutionary approach to geometry. He introduced a coordinate system that allowed geometric shapes to be represented by algebraic equations, thus laying the foundation for what is now known as analytic geometry. This breakthrough allowed mathematicians to solve geometric problems using algebraic methods and vice versa, revolutionizing the way mathematics was approached.

Descartes' coordinate system, now commonly referred to as the Cartesian coordinate system, consists of two perpendicular lines intersecting at a point called the origin. These lines, known as the x-axis and y-axis, enable the precise location of any point in a two-dimensional space. By assigning numerical values to these points,

Descartes was able to express geometric shapes using algebraic equations, thereby bridging the gap between geometry and algebra.

The impact of Descartes' work on analytic geometry cannot be overstated. It provided mathematicians with a powerful tool for solving complex geometric problems, paving the way for the development of calculus and other branches of modern mathematics. Moreover, Descartes' approach laid the groundwork for the application of mathematics in various scientific fields, such as physics and engineering.

Descartes' contributions to the development of analytic geometry showcase the power of interdisciplinary thinking and the ability to bridge seemingly disparate fields of study. His work not only transformed mathematics but also had a profound influence on the scientific community as a whole. Descartes' legacy as a pioneer of analytical thinking continues to resonate with students and scholars of intellectual history, reminding us of the profound impact that a single individual can have on the course of human knowledge.

Newton and the Mathematical Principles of Natural Philosophy

In the realm of intellectual history, few figures have left as profound an impact as Sir Isaac Newton. His groundbreaking work in the field of mathematics and physics revolutionized our understanding of the natural world and laid the foundation for modern science. In this subchapter, we will delve into Newton's remarkable achievements and explore the mathematical principles that underpin his discoveries.

Newton's magnum opus, "Mathematical Principles of Natural Philosophy," published in 1687, is a seminal work that transformed the scientific landscape. In this masterpiece, Newton introduced his three laws of motion, which served as the fundamental principles governing the movement of objects. These laws, known as Newton's Laws, provided a mathematical framework to explain and predict the behavior of bodies in motion, ranging from the motion of celestial bodies to everyday phenomena on Earth.

Central to Newton's Laws is the concept of force, which he defined as the product of an object's mass and its acceleration. This mathematical relationship between force, mass, and acceleration is now famously expressed in the equation $F = ma$. By quantifying and measuring forces, Newton was able to formulate a comprehensive theory of motion that contributed to the development of classical mechanics.

Another crucial aspect of Newton's work was his theory of universal gravitation. Inspired by the fall of an apple, Newton proposed that every particle in the universe attracts every other particle with a force proportional to their masses and inversely proportional to the square of the distance between them. This revolutionary theory enabled scientists to explain not only the motion of celestial bodies but also the tides, the orbits of planets, and the behavior of objects on Earth.

Newton's genius lay not only in his ability to formulate these mathematical principles but also in his application of them to solve complex problems. His invention of calculus, a mathematical tool that deals with rates of change and the calculation of infinitesimals, played a vital role in his investigations. By combining his laws of motion with calculus, Newton was able to solve intricate problems and make precise predictions about the behavior of physical systems.

In conclusion, Newton's "Mathematical Principles of Natural Philosophy" stands as a testament to his intellectual prowess and enduring legacy. His mathematical principles, including the laws of motion and the theory of universal gravitation, revolutionized the scientific world and continue to shape our understanding of the natural world today. As students of intellectual history, it is essential to recognize the profound impact of Newton's work and appreciate the mathematical foundations on which it stands.

Chapter 8: The Revolution in Biology and Botany

Linnaeus and the Classification of Species

In the vast realm of intellectual history, few individuals have made as significant a contribution as Carl Linnaeus to the understanding and classification of species. Linnaeus, a Swedish botanist and zoologist, is often referred to as the father of modern taxonomy. His groundbreaking work laid the foundation for our current understanding of the diversity and organization of life on Earth.

Born in 1707, Linnaeus dedicated his life to studying and categorizing nature's vast array of organisms. Prior to his work, there was no standardized system for naming and classifying species. Linnaeus realized the need for a coherent and consistent approach to organizing the vast number of species that existed, and set out to create a comprehensive system.

His most significant contribution was the development of binomial nomenclature, a naming system that assigns each species a unique two-part Latin name. This system, still in use today, consists of a genus name followed by a species name. For example, Homo sapiens is the scientific name for humans. Linnaeus believed that this standardized naming system would provide clarity and facilitate communication among scientists across different languages and cultures.

Linnaeus also developed a hierarchical classification system that organized species into a nested structure. He grouped related species into broader categories, such as genera, families, and orders, based on their shared characteristics. This hierarchical approach allowed for a better understanding of the relationships between species and provided a framework for further scientific exploration.

His most famous work, "Systema Naturae," published in 1735, laid out his classification system and described thousands of species. This monumental work not only provided a comprehensive catalog of known species but also introduced the concept of species being part of a larger natural order.

Linnaeus' classification system revolutionized the field of biology and had a profound impact on the study of natural history. It allowed scientists to organize and compare species, facilitating the discovery of new species and the identification of evolutionary relationships. Linnaeus' system also provided a framework for future scientific advancements, such as Charles Darwin's theory of evolution.

Today, Linnaeus' contributions continue to shape our understanding of the natural world. His work remains the foundation of modern taxonomy, and his legacy as a scientific pioneer endures. Students of intellectual history will find great value in studying Linnaeus' life and work, as it represents a crucial turning point in the scientific revolution and the development of our understanding of the natural world.

Darwin and the Theory of Evolution

Charles Darwin, a renowned English biologist, forever changed the way we understand the natural world with his groundbreaking theory of evolution. In this subchapter, we will delve into the life and work of Darwin, exploring how his ideas revolutionized intellectual history and continue to shape our understanding of the world today.

Born in 1809, Darwin grew up in a family of physicians and scientists, which fostered his curiosity about the natural world from an early age. After studying theology at Cambridge University, he embarked on a five-year journey aboard the HMS Beagle, which took him to various parts of the world, including the Galapagos Islands. This voyage provided Darwin with a wealth of data and observations that would later form the backbone of his theory.

Upon his return to England, Darwin meticulously analyzed his findings and began to develop his theory of evolution. He proposed that all life on Earth descended from a common ancestor and that species evolve through a process he called natural selection. According to Darwin, individuals with advantageous traits are more likely to survive and reproduce, passing those traits on to their offspring. Over time, this gradual accumulation of small changes leads to the formation of new species.

Darwin's theory of evolution challenged prevailing beliefs about the origins and diversity of life, particularly religious views that insisted on the divine creation of species. His groundbreaking book, "On the Origin of Species," published in 1859, presented a compelling argument for the theory of evolution, backed by extensive evidence and meticulous observations.

The impact of Darwin's theory on intellectual history cannot be overstated. It sparked intense debates among scientists, theologians, and philosophers, forcing them to reevaluate their understanding of the natural world and humanity's place within it. The theory of evolution also had profound implications for fields such as anthropology, genetics, and ecology, providing a framework for further scientific exploration.

Today, Darwin's theory of evolution remains a cornerstone of modern biology and continues to be refined and expanded upon by scientists around the world. It has helped us understand the diversity of life, the mechanisms of adaptation, and the interconnectedness of all living beings. Moreover, it has prompted us to critically examine our relationship with the natural world and the urgent need to preserve biodiversity.

In conclusion, Charles Darwin's theory of evolution stands as a monumental milestone in intellectual history. His meticulous observations, groundbreaking ideas, and compelling arguments have revolutionized our understanding of life on Earth. As students of intellectual history, it is important to appreciate the profound impact of Darwin's work and the ongoing legacy it carries in shaping our understanding of the natural world.

Mendel and the Discovery of Genetic Inheritance

In the realm of intellectual history, few figures have left as lasting an impact as Gregor Mendel. Known as the father of modern genetics, Mendel's groundbreaking work in the mid-19th century laid the foundation for our understanding of genetic inheritance, revolutionizing the field of biology and shaping our comprehension of the natural world.

Born in 1822 in what is now the Czech Republic, Mendel initially pursued a career in teaching, joining an Augustinian monastery. It was during this time that he began his pioneering experiments with pea plants, meticulously observing their traits and patterns of inheritance. His meticulous approach and attention to detail allowed him to make significant discoveries that would later become known as Mendel's Laws of Inheritance.

Mendel's experiments involved cross-breeding different varieties of pea plants, carefully tracking the traits exhibited by each generation. He observed that certain traits, such as seed color or flower color, would consistently appear in offspring in predictable ratios. This led him to propose the existence of "dominant" and "recessive" traits, where dominant traits would mask the expression of recessive traits in offspring.

One of Mendel's most significant contributions was his discovery of the principles of segregation and independent assortment. He observed that traits were inherited independently of each other, contradicting the prevailing belief that traits were blended together in offspring. This insight formed the basis for understanding genetic variation and inheritance, providing a framework for future scientists to build upon.

Unfortunately, Mendel's work went largely unnoticed during his lifetime. It wasn't until the early 20th century, decades after his death, that his groundbreaking research was rediscovered and widely recognized. His ideas laid the groundwork for the modern field of genetics and served as a catalyst for further scientific advancements, ultimately leading to the unraveling of the structure of DNA and the mapping of the human genome.

Mendel's contributions to intellectual history are immeasurable. His meticulous approach, combined with his keen observational skills, allowed him to unravel the secrets of genetic inheritance, forever changing our understanding of biology and the natural world. Today, his laws continue to be taught in biology classrooms around the world, inspiring future generations of scientists to explore the mysteries of life.

Chapter 9: The Impact and Legacy of the Scientific Revolution

The Enlightenment and the Spread of Scientific Ideas

Introduction:

The Enlightenment was a significant intellectual and cultural movement that took place in Europe during the 17th and 18th centuries. It marked a shift in thinking and a departure from traditional beliefs, paving the way for the spread of scientific ideas. This subchapter explores the impact of the Enlightenment on the dissemination of scientific knowledge, its key figures, and the resulting transformation in intellectual history.

The Enlightenment and Scientific Progress:

The Enlightenment emphasized reason, logic, and empirical evidence as the foundations for knowledge and understanding. This newfound emphasis on rational thinking led to a surge in scientific progress. Scholars and philosophers sought to understand the natural world through observation and experimentation, laying the groundwork for modern science.

Key Figures of the Enlightenment:

Prominent figures emerged during this period, who championed the spread of scientific ideas and played pivotal roles in shaping intellectual history. One such figure was Sir Isaac Newton, whose laws of motion and universal gravitation revolutionized physics. Newton's work not only advanced scientific knowledge but also laid the foundation for future scientific discoveries.

Another influential figure was Voltaire, a French philosopher, and writer. Voltaire advocated for religious tolerance, freedom of thought, and the importance of reason. His writings popularized scientific ideas and encouraged critical thinking among the masses, ultimately contributing to the Enlightenment's impact on intellectual history.

Spreading Scientific Ideas:

The Enlightenment era saw the rise of printing and publishing, which played a crucial role in disseminating scientific knowledge. Scientific journals and books became more accessible, allowing for the widespread sharing of ideas. This enabled scientists and intellectuals to communicate their discoveries and theories, fostering a vibrant exchange of knowledge.

The Impact on Intellectual History:

The Enlightenment and the spread of scientific ideas brought about a paradigm shift in intellectual history. It challenged traditional beliefs rooted in religious dogma and superstition, replacing them with evidence-based reasoning. This intellectual revolution laid the groundwork for the modern scientific method, emphasizing observation, experimentation, and logical analysis.

Conclusion:

The Enlightenment was a transformative period in intellectual history, characterized by the spread of scientific ideas and a shift towards reason and empirical evidence. Through the works of key figures like Newton and Voltaire, scientific knowledge became more accessible to a wider audience. This period marked a turning point in human understanding, paving the way for the advancements in science and technology that we benefit from today. By appreciating the Enlightenment's impact, students of intellectual history can gain a deeper understanding of the origins of the scientific revolution and its lasting influence on human progress.

The Industrial Revolution and Technological Advancements

The Industrial Revolution marked a pivotal turning point in human history, forever altering the course of society and setting in motion a chain of events that would shape the world we know today. This subchapter will delve into the intricacies of this transformative period, emphasizing the profound impact it had on technological advancements and intellectual history.

During the 18th and 19th centuries, a wave of industrialization swept across Europe and later spread to other parts of the world. This period witnessed a rapid shift from agrarian-based economies to industrialized ones, as new inventions and innovations revolutionized the way goods were produced. The introduction of steam power, mechanized textile production, and the development of the factory system ushered in an era of unprecedented productivity and economic growth.

One of the key factors driving this industrial boom was the surge in technological advancements. Inventors and engineers, fueled by a spirit of curiosity and ingenuity, brought forth a myriad of groundbreaking creations that propelled society into the modern age. The steam engine, for instance, was a game-changer. Its application to various industries, such as transportation and manufacturing, increased efficiency and productivity to levels previously unimaginable.

Furthermore, the Industrial Revolution had a profound impact on intellectual history. The rapid pace of technological advancements prompted scholars and thinkers to reflect upon the consequences of this newfound progress. Intellectual debates and philosophical discussions emerged, questioning the implications of industrialization on society, morality, and human progress. Writers like Karl Marx and Friedrich Engels, for instance, critiqued the exploitative nature of capitalism arising from the industrial system, advocating for a fairer distribution of wealth and an overhaul of societal structures.

Moreover, the Industrial Revolution led to the rise of the middle class and an increase in literacy rates, fostering a thirst for knowledge among individuals. As a result, educational institutions and libraries flourished, providing access to a wealth of information. This, in turn, spurred a greater interest in intellectual pursuits and the exploration of scientific theories, further propelling the scientific revolution.

In conclusion, the Industrial Revolution and its subsequent technological advancements irrevocably transformed society, leaving an indelible mark on intellectual history. Understanding the impact of this period is crucial for students of intellectual history as it provides invaluable insights into the forces that shaped our modern world. By examining the innovations and ideas that emerged during this era, we can better comprehend the profound consequences of the Industrial Revolution and its enduring legacy.

Modern Science: Building upon the Foundations

In the vast expanse of human history, the Scientific Revolution stands as a pivotal period that changed the course of intellectual development forever. This subchapter, titled "Modern Science: Building upon the Foundations," delves into the fascinating journey of how science has evolved over time, shaping the world we live in today. Addressed to students with an interest in intellectual history, this section will unravel the mysteries behind the growth and achievements of modern science.

To comprehend the significance of modern science, we must first acknowledge the foundations upon which it was built. The Scientific Revolution did not occur in isolation; it was rooted in centuries of intellectual inquiry and the accumulation of knowledge from diverse cultures across the globe. From the ancient Greeks to the Islamic Golden Age, each civilization made valuable contributions that laid the groundwork for future scientific advancements.

The subchapter explores the critical role played by individuals such as Copernicus, Galileo, and Newton in revolutionizing scientific thought. These remarkable minds challenged prevailing beliefs and developed groundbreaking theories that overturned long-established notions about the natural world. The transformation of astronomy, physics, and biology during this period opened up new avenues for inquiry and propelled humanity towards a more comprehensive understanding of the universe.

Furthermore, this section highlights the interconnectedness between scientific progress and societal changes. The emergence of the scientific method, a systematic approach to experimentation and observation, led to a shift in the way knowledge was acquired and validated. This methodical approach not only influenced scientific disciplines but also impacted broader fields such as politics, philosophy, and economics.

As students of intellectual history, it is crucial to recognize that modern science is an ongoing process, continuously expanding upon the foundations laid by our predecessors. This subchapter will explore the subsequent waves of scientific development, including the Industrial Revolution, the Information Age, and the current era of technological advancements. By understanding the historical context of these scientific breakthroughs, students can gain a deeper appreciation for the discoveries and innovations that shape our lives today.

In conclusion, "Modern Science: Building upon the Foundations" is a subchapter that delves into the evolution of scientific thought, tracing its roots back to ancient civilizations and highlighting the contributions of remarkable individuals. By exploring the interconnectedness between scientific progress and broader societal changes, students with an interest in intellectual history can gain a comprehensive understanding of the transformative power of modern science.

Chapter 10: The Unresolved Questions and Future Directions

The Nature of Dark Matter and Dark Energy

In the vast expanse of the universe, there exist mysterious forces that continue to baffle scientists and ignite our curiosity. Dark matter and dark energy, although invisible to the naked eye, play a crucial role in shaping the cosmos as we know it. In this subchapter, we will delve into the enigmatic nature of these phenomena, shedding light on their significance and the ongoing scientific quest to unravel their mysteries.

Dark matter, as the name suggests, refers to a form of matter that does not interact with light or any other electromagnetic radiation. Its existence was first proposed by Swiss astronomer Fritz Zwicky in the 1930s when he observed that the gravitational pull on galaxies was far greater than could be accounted for by visible matter alone. Since then, extensive research and observations have confirmed the presence of dark matter, which outweighs visible matter by a staggering margin.

So, what is dark matter made of? Well, that remains a mystery. Scientists have put forth several theories, ranging from weakly interacting massive particles (WIMPs) to axions and primordial black holes. However, until now, direct detection of dark matter particles has eluded us. This is where the astute minds of the intellectual history come into play, tirelessly working to uncover the secrets hidden within the cosmos.

On the other hand, dark energy is an even more elusive concept. Discovered relatively recently, in the late 1990s, dark energy is believed to be responsible for the accelerated expansion of the universe. Unlike dark matter, which exerts gravitational force, dark energy possesses negative pressure, pushing galaxies apart at an ever-increasing rate. This discovery came as a surprise and revolutionized our understanding of the cosmos, leading to the 2011 Nobel Prize in Physics for the scientists who made these breakthrough observations.

The study of dark matter and dark energy has significant implications for our knowledge of the universe's past, present, and future. Understanding their nature will help us comprehend the fate of the cosmos, whether it will continue expanding indefinitely or meet a different destiny. Moreover, the pursuit of knowledge in this field has deepened our understanding of gravity, particle physics, and the fundamental laws that govern our universe.

As students of the scientific revolution, we have the privilege of witnessing the ongoing discoveries and advancements in these intellectual frontiers. The quest to unravel the nature of dark matter and dark energy offers an exciting opportunity to contribute to humanity's collective understanding of the universe. So, let us embark on this journey with open minds and unyielding curiosity, as we delve into the mysteries that lie beyond the visible cosmos.

The Search for Extraterrestrial Life

In the vast expanse of the universe, spanning billions of light-years, humanity has always been intrigued by the possibility of extraterrestrial life. The quest to unravel the mysteries of the cosmos and discover if we are truly alone has captivated the minds of scientists, philosophers, and dreamers alike. This chapter delves into the fascinating field of astrobiology and the tireless efforts made by humanity in the search for extraterrestrial life.

The search for extraterrestrial life is not a recent phenomenon but has deep roots in our intellectual history. From ancient civilizations pondering the existence of otherworldly beings to the groundbreaking discoveries of the scientific revolution, the quest for understanding our place in the universe has been a driving force behind human progress. This subchapter explores the historical context and the pivotal moments that have shaped our search for extraterrestrial life.

We begin by exploring ancient myths and legends that revolve around encounters with beings from other planets. These tales, passed down through generations, reflect humanity's innate curiosity about the possibility of life beyond Earth. Moving forward, we delve into the scientific revolution and its impact on our understanding of the universe. Visionaries like Galileo Galilei and Johannes Kepler paved the way for a more systematic approach to studying the cosmos, challenging traditional beliefs and opening up new possibilities for the existence of extraterrestrial life.

The subchapter also focuses on the modern era, where technological advancements have allowed scientists to explore distant planets and moons in our own solar system and beyond. The development of powerful telescopes and space probes has revolutionized our understanding of the universe, revealing the potential for habitable environments on planets like Mars and moons like Europa.

Furthermore, the subchapter will discuss the ongoing scientific missions, such as the Kepler Space Telescope and the Mars rovers, that have been instrumental in gathering data and searching for signs of life beyond Earth. It will also delve into the fascinating field of astrobiology, which combines various scientific disciplines to study the conditions necessary for life to exist elsewhere in the universe.

For the intellectual history niche, this subchapter will also examine the philosophical and ethical implications of discovering extraterrestrial life. How would such a breakthrough affect our understanding of religion, the nature of humanity, and our place in the cosmos? These questions have captivated the minds of philosophers and theologians for centuries, and we will explore the various perspectives on this profound topic.

In conclusion, "The Search for Extraterrestrial Life" subchapter in "The Scientific Revolution: Unveiling the Secrets of the Past" takes students on a captivating journey through history, science, and philosophy. It offers a comprehensive overview of humanity's quest to uncover the truth about life beyond Earth, showcasing the remarkable progress made and the thought-provoking questions that arise along the way.

The Frontiers of Quantum Mechanics and String Theory

In the vast realm of scientific exploration, few areas captivate the imagination and challenge the limits of our understanding like quantum mechanics and string theory. Delving into the inner workings of the universe, these frontiers of modern physics have revolutionized our perception of reality and reshaped the course of intellectual history. In this subchapter, we embark on a journey to uncover the secrets behind these groundbreaking theories, aiming to provide students and enthusiasts of intellectual history with a glimpse into the awe-inspiring world of quantum mechanics and string theory.

Quantum mechanics, the fundamental theory of the microscopic world, defies our classical understanding of physics. It reveals a realm where particles can exist in multiple states simultaneously, where measurements can alter the very properties they seek to observe, and where uncertainty reigns supreme. The pioneers of quantum mechanics, such as Max Planck, Albert Einstein, and Niels Bohr, challenged the prevailing Newtonian paradigm and paved the way for a new era of scientific thought. Students of intellectual history will appreciate the intellectual and philosophical upheaval caused by these revolutionary ideas, as they challenged the deterministic worldview that had prevailed for centuries.

String theory, on the other hand, pushes the boundaries of quantum mechanics even further. It proposes that at the most fundamental level, all matter and energy in the universe are composed of tiny, vibrating strings. These strings, whose vibrations determine the properties of particles, offer a tantalizing glimpse of a unified theory that can reconcile the seemingly incompatible theories of quantum mechanics and general relativity. While still a work in progress, string theory has captivated the minds of physicists and intellectuals alike, offering new insights into the nature of space, time, and the cosmos.

As students of intellectual history delve into the frontiers of quantum mechanics and string theory, they will encounter not only the scientific concepts but also the profound implications for our understanding of reality. These theories challenge our conventional notions of cause and effect, time, and the very fabric of the universe. They raise profound questions about the nature of existence, consciousness, and the limits of human knowledge. Exploring these frontiers provides an opportunity to witness the ongoing dialogue between science and philosophy, as scientists grapple with the mind-boggling implications of these theories.

In conclusion, the frontiers of quantum mechanics and string theory offer a captivating journey for students and intellectual history enthusiasts alike. By delving into these theories, one not only gains a deeper understanding of the nature of the universe but also witnesses the transformative power of scientific revolutions. From the conceptual revolutions brought forth by quantum mechanics to the tantalizing possibilities offered by string theory, these frontiers have forever altered our perception of reality and continue to shape the course of intellectual history.

Printed in the USA
CPSIA information can be obtained
at www.ICGtesting.com
CBHW070319030724
11009CB00019B/1354